Elevating Excellence: The Top 10 Essentials of Aviation Maintenance

Amman Mujahid

ISBN: 9798334138315

DEDICATION

Dedicated to my father, mother, sisters and husband.

CONTENTS

	Acknowledgments	i
1	Chapter 1: Top 10 Aircraft Maintenance Checks	Pg#2
2	Chapter 2: Top 10 Preventative Maintenance Practices	Pg #4
3	Chapter 3: Top 10 Tips for Effective Troubleshooting	Pg #6
4	Chapter4: Top 10 Most Commonly Replaced Aircraft Parts	Pg #8
5	Chapter 5: Top 10 Aviation Maintenance Training Programs	Pg #10
6	Chapter 6: Top 10 Aircraft Maintenance Tools	Pg #12
7	Chapter 7: Top 10 Aircraft Maintenance Software Solutions	Pg #14
8	Chapter 8: Top 10 Environmental Considerations in Aviation Maintenance	Pg #16
9	Chapter 9: Top 10 Innovations in Aircraft Materials	Pg #18
10	Chapter 10: Conclusion	Pg #20

Introduction

I would not say I was always a fan of aircrafts and aviation until I chose to dive into the field of aviation maintenance. Realizing that innovation not only lies in building and manufacturing machines but maintaining them too. Being a student of aviation maintenance engineering, I always rooted for more content that could resonate with my layman mind to have an in depth and easy understanding of the career I wanted to be a part of. As the field itself was never conventional, getting hands on such information was always a challenge. This work of mine may not be able to fill in the gap of knowledge delivery but will for sure lead to a stepping stone within the process.

CHAPTER 1: TOP 10 AIRCRAFT MAINTENANCE CHECKS

1. Daily Checks
Description: Routine inspections performed daily to ensure the aircraft is operational and ready for flight. Includes checking fluid levels, tire pressures, and visual inspections for any signs of damage or wear.

2. Pre-flight Inspections
Description: Comprehensive checks conducted before each flight to verify that the aircraft is in optimal condition. Involves inspecting critical systems and components such as engines, avionics, and control surfaces to ensure readiness and safety.

3. Post-flight Inspections
Description: Inspections carried out after each flight to identify any issues that may have arisen during operation. Focuses on examining engine performance, landing gear, and identifying any unusual noises or vibrations.

4. A-Checks (Line Maintenance)
Description: Routine maintenance tasks performed at regular intervals to address minor issues and ensure operational efficiency. Includes visual inspections, functional checks, and minor repairs of aircraft systems and components.

5. B-Checks (Intermediate Maintenance)
Description: More detailed inspections and maintenance tasks than A-checks, addressing system reliability and performance. Involves deeper examinations and component replacements to ensure continued airworthiness.

6. C-Checks (Heavy Maintenance)
Description: Comprehensive inspections involving extensive disassembly and examination of major aircraft components. Typically performed at longer intervals, C-checks focus on in-depth inspections, repairs, and overhauls.

7. D-Checks (Complete Overhaul)
Description: The most extensive and thorough maintenance check, involving complete disassembly of the aircraft. Includes detailed inspections, overhauls, and refurbishments to ensure the aircraft's long-term safety and performance.

8. Special Inspections (After Unusual Occurrences)

Description: Inspections conducted in response to specific incidents or unusual occurrences, such as hard landings or system anomalies. Focuses on diagnosing and addressing any issues resulting from these events.

9. Non-Destructive Testing (NDT) Techniques

Description: Specialized testing methods used to detect internal or hidden defects without damaging the aircraft. Common techniques include ultrasonic, radiographic, and eddy current testing to ensure structural integrity.

10. Corrosion Prevention and Control Program (CPCP) Checks

Description: Procedures focused on preventing and managing corrosion, which can significantly impact aircraft safety and longevity. Involves regular inspections, treatment, and preventive measures to address and control corrosion.

Chapter 2: Top 10 Preventative Maintenance Practices

In the realm of aviation maintenance, proactive measures are key to ensuring safety, reliability, and performance. **Chapter 2: Top 10 Preventative Maintenance Practices** delves into essential practices designed to pre-emptively address potential issues before they impact aircraft operations.

1. Regular Lubrication

Description: Consistent application of lubrication to moving parts reduces friction and wear, thereby extending the lifespan of critical components. This practice helps maintain smooth operation and prevents premature failure.

2. Scheduled Engine Runs

Description: Routine engine runs are performed to ensure optimal engine performance and identify any anomalies early. These tests help in monitoring engine health, verifying functionality, and preventing unexpected failures.

3. Tire Pressure Monitoring

Description: Regular monitoring and adjustment of tire pressure are crucial for safe aircraft operations. Proper tire pressure ensures efficient fuel consumption, smooth landings, and reduces the risk of tire blowouts.

4. Fluid Level Checks

Description: Regular checks of essential fluids—such as hydraulic fluids, oil, and coolant—ensure that systems operate efficiently and prevent potential malfunctions due to fluid shortages or contamination.

5. Component Replacement Before Failure

Description: Pre-emptive replacement of parts based on usage cycles or wear limits prevents unexpected failures and costly repairs. This strategy ensures that components are replaced before reaching the end of their useful life.

6. Routine Cleaning and De-greasing

Description: Regular cleaning and de-greasing of aircraft components prevent buildup that can lead to performance issues or corrosion. This practice maintains operational efficiency and extends the life of the equipment.

7. Battery Health Checks

Description: Periodic assessment of battery health ensures reliable power supply to critical systems. Monitoring battery charge, capacity, and connections helps avoid sudden power failures and operational disruptions.

8. Flight Control Surface Lubrication

Description: Proper lubrication of flight control surfaces reduces friction and wear, ensuring smooth and responsive control. This practice is essential for maintaining precise handling and overall aircraft performance.

9. Regular Calibration of Instruments

Description: Ensuring that aircraft instruments are regularly calibrated maintains their accuracy and reliability. Proper calibration is vital for accurate readings and safe navigation during flight operations.

10. Scheduled Software Updates for Avionics

Description: Routine updates to avionics software enhance functionality, fix bugs, and improve system performance. Keeping software current ensures that avionics systems operate effectively and integrate with the latest technological advancements.

Chapter 3: Top 10 Tips for Effective Troubleshooting

In the high-stakes world of aviation maintenance, effective troubleshooting is essential for quickly diagnosing and resolving issues that could impact aircraft safety and performance. This chapter offers a comprehensive guide to mastering the art of problem-solving in aviation maintenance. This chapter provides critical insights and practical strategies to enhance your troubleshooting skills and ensure precise and efficient resolution of technical problems.

1. Understand the Aircraft Systems

- **Description:** A deep understanding of the aircraft's systems and their interactions is foundational for effective troubleshooting. Knowledge of how systems operate together helps in accurately identifying where problems may originate and how they affect overall performance.

2. Use the Aircraft Maintenance Manual (AMM)

- **Description:** The AMM is an invaluable resource for detailed procedures and specifications. Leveraging this manual ensures adherence to manufacturer guidelines, facilitates correct diagnostic steps, and enhances accuracy in maintenance tasks.

3. Start with Simple Checks

- **Description:** Begin troubleshooting with basic checks to eliminate common and straightforward issues. This methodical approach can quickly identify minor problems and avoid unnecessary complexity in the troubleshooting process.

4. Utilize Diagnostic Tools

- **Description:** Employing specialized diagnostic tools and equipment—such as multimeters, borescopes, and data readers—provides precise measurements and insights into system performance. These tools are crucial for detecting faults that are not immediately visible.

5. Check Maintenance Records

- **Description:** Reviewing historical maintenance records can provide valuable context for current issues. Analyzing past repairs, replacements, and recurring problems can help pinpoint underlying issues and inform troubleshooting strategies.

6. Isolate the Problem

- **Description:** Systematically isolating the problem involves narrowing down the potential sources of the issue. This focused approach helps in identifying the exact component or system that is malfunctioning, streamlining the troubleshooting process.

7. Test Hypotheses Methodically
- **Description:** Formulating and testing hypotheses in a structured manner allows for systematic problem-solving. This approach involves making educated guesses about the cause of the issue and conducting tests to confirm or refute these theories.

8. Consult with Experienced Colleagues
- **Description:** Collaboration with experienced colleagues can provide new perspectives and insights. Consulting with others who have faced similar issues can offer valuable advice and enhance the effectiveness of the troubleshooting process.

9. Verify Repairs Thoroughly
- **Description:** After making repairs, it is crucial to thoroughly verify that the issue has been resolved. Comprehensive testing ensures that the repair is effective and that no additional problems have been introduced.

10. Document Findings and Actions Taken
- **Description:** Accurate documentation of findings and actions is essential for future reference and continuous improvement. Detailed records provide a valuable reference for future troubleshooting and contribute to overall maintenance management.

Chapter 4: Top 10 Most Commonly Replaced Aircraft Parts

Certain components undergo frequent replacement due to their critical roles and wear-and-tear during operation. **Chapter 4: Top 10 Most Commonly Replaced Aircraft Parts** provides an in-depth exploration of these vital components, shedding light on their importance, replacement criteria, and the impact of their maintenance on overall aircraft performance and safety.

1. Tires

Description: Aircraft tires are subjected to substantial stress during takeoffs and landings. Regular replacement ensures optimal performance and safety, as well as prevents issues such as blowouts and uneven wear that can affect flight stability.

2. Brake Pads

Description: Brake pads are crucial for safe landing and ground maneuvering. Replacing worn brake pads maintains effective braking performance and prevents potential failures that could compromise aircraft safety.

3. Spark Plugs

Description: Spark plugs are essential for igniting the fuel-air mixture in aircraft engines. Regular replacement ensures consistent engine performance, fuel efficiency, and reliable starting.

4. Fuel Filters

Description: Fuel filters are vital for removing contaminants from the fuel system, ensuring clean fuel delivery to the engines. Regular replacement helps prevent engine clogging and performance issues caused by impurities.

5. Oil Filters

Description: Oil filters keep the engine oil clean by removing debris and contaminants. Replacing oil filters at recommended intervals helps maintain engine efficiency, prevent damage, and extend engine life.

6. Batteries

Description: Aircraft batteries provide essential power for engine starts and various systems. Regular replacement of batteries ensures reliable electrical power and minimizes the risk of sudden failures.

7. Windshields

Description: Windshields protect pilots and crew from environmental elements and impact. Replacing damaged or worn windshields maintains visibility and ensures the structural integrity of the cockpit.

8. Light Bulbs

Description: Light bulbs are used in various aircraft systems for visibility and signalling. Regular replacement ensures proper illumination, enhancing safety during operations and night flights.

9. O-rings and Seals

Description: O-rings and seals prevent leaks in fluid systems and maintain pressure. Replacing these components as needed ensures system integrity and prevents fluid leaks that could lead to system failures.

10. Fan Blades

Description: Fan blades, critical in engine cooling and airflow, are subject to wear and damage. Regular replacement of fan blades ensures efficient engine cooling and prevents potential overheating.

Chapter 5: Top 10 Aviation Maintenance Training Programs

In the dynamic and highly specialized field of aviation maintenance, staying current with industry standards and technological advancements is essential. **Chapter 5: Top 10 Aviation Maintenance Training Programs** delves into the most impactful training programs designed to enhance the skills and knowledge of aviation maintenance professionals. This chapter highlights top training opportunities that not only equip technicians with the latest techniques but also ensure compliance with global standards, ultimately fostering a culture of excellence in aviation maintenance.

1. FAA-Approved Part 147 Schools
Description: FAA-Approved Part 147 schools offer foundational and advanced training tailored to meet Federal Aviation Administration requirements. These programs provide critical skills and certification necessary for a successful career in U.S. aviation maintenance.

2. EASA Part 66 Training Programs
Description: EASA Part 66 training programs align with European Union Aviation Safety Agency regulations, offering in-depth knowledge for technicians working within Europe. These courses cover comprehensive maintenance practices and ensure adherence to EU standards.

3. OEM Training Courses
Description: OEM training courses are designed to deliver specialized education on specific aircraft and component systems from original equipment manufacturers. This training ensures technicians are proficient in the latest technologies and manufacturer protocols.

4. Avionics Maintenance Training
Description: Focused on the intricate world of aircraft electronics, avionics maintenance training covers the repair and upkeep of advanced navigation and communication systems. This training is critical for maintaining the sophisticated electronics that modern aircraft rely on.

5. Composite Repair Courses
Description: As aircraft materials evolve, composite repair courses become increasingly vital. These programs teach the repair techniques for advanced composite materials, ensuring that technicians can address issues related to modern aircraft designs.

6. NDT Certification Programs

Description: Non-Destructive Testing (NDT) certification programs offer essential training for inspecting aircraft components without causing damage. This specialized training is key for detecting structural issues and ensuring safety.

7. Engine-Specific Training

Description: Engine-specific training provides targeted instruction on the maintenance of particular engine types. This program is crucial for mastering the complexities of various engine models and ensuring optimal engine performance.

8. Online Refresher Courses

Description: Online refresher courses offer flexible, up-to-date learning options for aviation maintenance professionals. These courses are perfect for staying current with industry advancements and regulatory changes while managing busy schedules.

9. Human Factors in Aviation Maintenance

Description: Training in human factors addresses the role of human behaviour in maintenance processes. This course emphasizes error prevention, safety protocols, and effective communication to enhance overall maintenance quality and reduce mistakes.

10. Safety Management Systems (SMS) Training

Description: Safety Management Systems (SMS) training focuses on implementing robust safety protocols within aviation operations. This program equips professionals with the knowledge to develop and manage safety systems that mitigate risks and promote a safety-oriented culture

Chapter 6: Top 10 Aircraft Maintenance Tools

In aircraft maintenance, precision and reliability are paramount, and having the right tools at your disposal is essential for ensuring both. **Chapter 6** offers a comprehensive guide to the most critical instruments used by aviation professionals to maintain and repair aircraft efficiently and accurately. This chapter delves into each tool's specific applications, highlighting their importance in ensuring the safety, performance, and longevity of aircraft.

1. Borescope

Description: A borescope is an essential tool for inspecting internal components of an aircraft engine or other hard-to-reach areas without disassembly. Its ability to provide real-time, high-resolution images helps technicians identify potential issues and assess the condition of internal parts.

2. Torque Wrench

Description: The torque wrench is crucial for applying precise amounts of torque to bolts and fasteners. Accurate torque settings are vital for ensuring the structural integrity and safety of aircraft assemblies.

3. Multimeter

Description: A multimeter measures various electrical parameters such as voltage, current, and resistance. This tool is indispensable for diagnosing electrical issues, ensuring proper functionality of aircraft systems, and troubleshooting complex electrical circuits.

4. Safety Wire Pliers

Description: Safety wire pliers are used to secure bolts and other fasteners with safety wire, preventing them from loosening during operation. This tool is crucial for maintaining the integrity of critical aircraft components and preventing potential failures.

5. Ultrasonic Tester

Description: An ultrasonic tester is used for detecting internal flaws and measuring the thickness of materials through ultrasonic waves. This tool helps in identifying issues such as corrosion or cracks that are not visible to the naked eye.

6. Rivet Gun

Description: The rivet gun is essential for fastening metal components together using rivets. This tool is commonly used in aircraft assembly and repair, providing a strong, reliable connection between parts.

7. Aircraft Jack

Description: Aircraft jacks are used to lift and support an aircraft during maintenance procedures. They are vital for accessing the undercarriage and performing tasks such as tire changes, brake inspections, and landing gear maintenance.

8. Precision Calipers

Description: Precision calipers measure the dimensions of components with high accuracy. They are essential for ensuring that parts meet exact specifications and tolerances required for safe and effective operation.

9. Magneto Timing Light

Description: The magneto timing light is used to set the timing of the ignition system in aircraft engines. Proper timing is critical for engine performance, fuel efficiency, and overall reliability.

10. Infrared Thermometer

Description: An infrared thermometer measures surface temperatures without direct contact. It is useful for diagnosing overheating components, checking engine performance, and ensuring that systems operate within safe temperature ranges.

Chapter 7: Top 10 Aircraft Maintenance Software Solutions

In the high-stakes world of aviation maintenance, leveraging advanced software solutions is key to optimizing operations, ensuring compliance, and enhancing efficiency. Let us explore the leading digital tools designed to streamline maintenance processes, manage aircraft records, and support decision-making. This chapter provides an in-depth look at each software solution, highlighting its features, benefits, and how it contributes to effective maintenance management.

1. CAMP Systems
Description: CAMP Systems offers a comprehensive aircraft maintenance tracking and management solution. Its user-friendly interface and robust features support compliance, documentation, and reporting, making it a top choice for managing aircraft maintenance programs.

2. Traxxall
Description: Traxxall is a cloud-based maintenance management system that provides real-time tracking and reporting. It enables users to monitor aircraft health, manage maintenance schedules, and ensure regulatory compliance with ease.

3. Flightdocs
Description: Flightdocs delivers a powerful suite of tools for aircraft maintenance and inventory management. Its integrated platform helps streamline workflows, enhance record-keeping, and improve operational efficiency through real-time data access.

4. Quantum Control
Description: Quantum Control offers a versatile solution for managing aviation maintenance, parts inventory, and financial transactions. Its modular design allows for customization, providing tailored solutions for different maintenance and operations needs.

5. AMOS (Aircraft Maintenance and Engineering System)
Description: AMOS is an advanced software system used by major airlines and maintenance organizations. It provides extensive capabilities for maintenance planning, engineering, and compliance management, supporting complex maintenance activities and enhancing operational efficiency.

6. Rusada Envision
Description: Rusada Envision is a comprehensive aviation management software that covers maintenance, engineering, and inventory management. Its intuitive design and powerful analytics help organizations optimize maintenance processes and improve decision-making.

7. AvPro Software

Description: AvPro Software offers an integrated maintenance management solution with features for tracking maintenance, managing parts, and generating reports. It is designed to enhance visibility and control over maintenance operations.

8. Airworthiness Directives Tracking Systems

Description: This software focuses on managing and tracking airworthiness directives (ADs) to ensure compliance with regulatory requirements. It helps maintenance teams stay updated on safety directives and schedule necessary inspections or repairs.

9. Maintenance Performance Toolbox by Boeing

Description: The Maintenance Performance Toolbox by Boeing provides a suite of tools for maintenance planning, execution, and analysis. It supports efficient maintenance management through advanced data analytics and operational insights.

10. IFS Aviation and Défense Maintenance Software

Description: IFS Aviation and defence Maintenance Software offers a comprehensive solution for managing maintenance, supply chain, and engineering functions. Its advanced features support complex maintenance operations and enhance overall efficiency.

Chapter 8: Top 10 Environmental Considerations in Aviation Maintenance

As the aviation industry faces increasing scrutiny regarding its environmental impact, incorporating sustainable practices into aircraft maintenance becomes essential. This chapter offers actionable insights into how maintenance professionals can contribute to a greener aviation industry through effective environmental management.

1. Proper Disposal of Hazardous Materials

Description: Ensuring the safe and compliant disposal of hazardous materials, such as solvents and chemicals, is critical for minimizing environmental contamination. This section covers best practices for handling and disposing of these materials responsibly.

2. Use of Eco-Friendly Cleaning Products

Description: Eco-friendly cleaning products reduce the environmental impact of maintenance operations. This section highlights products that are both effective and environmentally safe, promoting sustainability while maintaining high cleaning standards.

3. Recycling of Aircraft Components

Description: Recycling aircraft components helps conserve resources and reduce waste. This section discusses methods for recycling parts and materials, including partnerships with recycling facilities and the benefits of material reclamation.

4. Energy-Efficient Hangar Operations

Description: Implementing energy-efficient practices in hangar operations can significantly reduce energy consumption. This section explores techniques for optimizing lighting, heating, and cooling systems to enhance energy efficiency.

5. Compliance with Environmental Regulations

Description: Adhering to environmental regulations is essential for legal compliance and reducing ecological impact. This section provides an overview of key regulations and guidelines that maintenance facilities must follow to ensure environmental responsibility.

6. Reducing VOC Emissions

Description: Volatile Organic Compounds (VOCs) contribute to air pollution and health hazards. This section examines strategies for reducing VOC emissions through the use of low-VOC products and improved ventilation systems.

7. Water Conservation Measures

Description: Water conservation is a critical aspect of environmental stewardship. This section covers techniques for minimizing water usage in maintenance activities, including the implementation of water-saving technologies and practices.

8. Noise Pollution Control

Description: Managing noise pollution from maintenance operations helps reduce the impact on surrounding communities. This section provides guidelines for controlling and mitigating noise through equipment modifications and operational practices.

9. Sustainable Material Use

Description: Using sustainable materials in maintenance activities helps reduce the environmental impact of repairs and replacements. This section discusses the benefits of choosing eco-friendly materials and the impact on overall sustainability.

10. Green Certifications for Maintenance Facilities

Description: Achieving green certifications demonstrates a commitment to environmental best practices. This section outlines the process for obtaining certifications and the advantages of being recognized as a green facility.

Chapter 9: Top 10 Innovations in Aircraft Materials

In the quest for more efficient, durable, and lightweight aircraft, material science plays a pivotal role. **Chapter 9: Top 10 Innovations in Aircraft Materials** delves into the cutting-edge materials that are transforming the aviation industry. This chapter provides an overview of the most revolutionary materials currently driving advancements in aircraft design, performance, and safety.

1. Carbon Fiber Composites

Description: Carbon fiber composites are renowned for their exceptional strength-to-weight ratio and rigidity. This section explores how these materials contribute to reducing aircraft weight and enhancing fuel efficiency while maintaining structural integrity.

2. Titanium Alloys

Description: Titanium alloys offer remarkable strength and corrosion resistance, making them ideal for high-stress components. This section examines their applications in aerospace and how they improve performance and durability.

3. Advanced Aluminum Alloys

Description: Advanced aluminum alloys combine lightweight properties with increased strength and fatigue resistance. This section discusses their role in reducing aircraft weight and improving overall efficiency.

4. Ceramic Matrix Composites

Description: Ceramic matrix composites provide high-temperature resistance and durability. This section highlights their use in engine components and braking systems, enhancing performance under extreme conditions.

5. Graphene

Description: Graphene, a single layer of carbon atoms arranged in a hexagonal lattice, offers exceptional electrical and thermal conductivity along with remarkable strength. This section explores its potential applications in improving aircraft performance and safety.

6. Shape Memory Alloys

Description: Shape memory alloys can return to their original shape after deformation when subjected to heat. This section details their innovative uses in actuators and other components, contributing to more reliable and adaptive systems.

7. Self-Healing Materials
Description: Self-healing materials have the ability to repair themselves after damage. This section investigates their use in enhancing the longevity and safety of aircraft structures by automatically addressing minor damages.

8. Superalloys
Description: Superalloys are designed to withstand extreme temperatures and stress. This section covers their critical role in turbine engines and other high-performance components, ensuring reliability and efficiency.

9. Nano-Coatings
Description: Nano-coatings provide protective layers that are both ultra-thin and highly effective against corrosion, wear, and environmental damage. This section explores how they enhance the lifespan and performance of various aircraft components.

10. Thermoplastic Composites
Description: Thermoplastic composites offer ease of processing and recycling, along with high strength and impact resistance. This section discusses their benefits in producing lightweight and durable aircraft parts.

Chapter 10: Top 10 Resources for Aviation Maintenance Professionals

In the fast-evolving field of aviation maintenance, staying informed and connected is crucial for career development and operational excellence. **Chapter 10: Top 10 Resources for Aviation Maintenance Professionals** provides a curated list of essential resources that support ongoing learning, industry engagement, and professional growth. This chapter helps maintenance professionals access the tools and knowledge they need to stay at the forefront of their field.

1. FAA Advisory Circulars

Description: FAA Advisory Circulars offer essential guidance on regulatory requirements and best practices in aviation maintenance. This section outlines how these documents provide valuable insights for ensuring compliance and maintaining high standards.

2. EASA Safety Publications

Description: EASA Safety Publications provide crucial information on safety standards and procedures within the European aviation sector. This section explains their role in promoting safety and enhancing maintenance practices across Europe.

3. OEM Maintenance Manuals

Description: Original Equipment Manufacturer (OEM) Maintenance Manuals are detailed guides on maintaining specific aircraft models. This section highlights their importance for accurate and effective maintenance, troubleshooting, and repairs.

4. Aviation Maintenance Magazines

Description: Aviation maintenance magazines offer up-to-date industry news, technical articles, and expert opinions. This section discusses how these publications keep professionals informed about the latest trends and innovations.

5. Industry Conferences and Expos

Description: Attending industry conferences and expos provides valuable networking opportunities and exposure to new technologies. This section details how these events offer insights into current industry developments and future trends.

6. Online Forums and Communities

Description: Online forums and communities offer platforms for professionals to share knowledge, seek advice, and discuss challenges. This section covers the benefits of participating in these virtual networks for peer support and information exchange.

7. Professional Organizations (e.g., ATEC, AMFA)
Description: Membership in professional organizations provides access to a wealth of resources, including training, certification, and advocacy. This section explores how joining organizations like ATEC and AMFA supports professional development and industry engagement.

8. Technical Training Schools
Description: Technical training schools offer specialized education and certification programs in aviation maintenance. This section highlights their role in providing hands-on training and advancing skills for career progression.

9. Aviation Maintenance Blogs
Description: Aviation maintenance blogs feature articles, case studies, and expert opinions on various maintenance topics. This section discusses how following these blogs helps professionals stay updated with practical advice and industry insights.

10. Webinars and Online Courses
Description: Webinars and online courses provide flexible learning opportunities on a range of aviation maintenance topics. This section outlines how these digital resources facilitate ongoing education and skill enhancement.

Conclusion

As we reach the final chapter of this journey, it is evident that the aviation industry stands at a crossroads, brimming with opportunities and challenges that will shape the future of travel, commerce, and human connection. Throughout this book, we have explored the fascinating world of aviation through the lens of STEM.

As we conclude, I invite you to act. Whether you are a student, educator, industry professional, or enthusiast, there are countless ways to contribute to the future of aviation. Volunteer your time, engage with your community, and continue to learn and innovate. Together, we can inspire the next generation of STEM professionals and ensure that the aviation industry continues to soar to new heights.

The sky is not the limit; it is just the beginning. With passion, education, and a commitment to innovation, the future of aviation is boundless. Let us embrace this journey with enthusiasm and determination, knowing that each step we take brings us closer to a brighter, more connected world.

Thank you for embarking on this journey with me. Here's to a future where the possibilities are as vast as the sky

Bibliography

Federal Aviation Administration (FAA), 2024. Aviation Maintenance Technician Handbook. [online] Available at: https://www.faa.gov/regulations_policies/handbooks_manuals/aircraft/amt_handbook/

Federal Aviation Administration (FAA), 2024. Human Factors in Aviation Maintenance. [online] Available at: https://www.faa.gov/about/initiatives/maintenance_hf/library/documents/media/human_factors_maintenance/hf-maintenance-2.pdf

European Aviation Safety Agency (EASA), 2023. Continuing Professional Development in Aviation Maintenance. [online] Available at: https://www.easa.europa.eu/document-library/continuing-professional-development.

Jorgensen, J., 2022. Understanding Aircraft Maintenance Issues. 3rd ed. Aviation Publications. Available at: https://www.aviationpublications.com/aircraft-maintenance-issues

International Air Transport Association (IATA), 2024. *Technology and Innovation in Aviation Maintenance*. [online] Available at: https://www.iata.org/en/pressroom/2024-releases/2024-01-15-01/

Federal Aviation Administration (FAA), 2023. *Aviation Maintenance Technician Handbook*. [online] Available at: https://www.faa.gov/regulations_policies/handbooks_manuals/aircraft/amt_handbook/